YOUR KNOWLEDGE HAS VALUE

- We will publish your bachelor's and master's thesis, essays and papers

- Your own eBook and book - sold worldwide in all relevant shops

- Earn money with each sale

Upload your text at www.GRIN.com and publish for free

Imprint:

Copyright © 2018 GRIN Verlag
Print and binding: Books on Demand GmbH, Norderstedt Germany
ISBN: 9783668722040

Austin Mwana

Heat Exchanger Failure

Investigation Report

GRIN Verlag

HEAT EXCHANGER FAILURE INVESTIGATION REPORT

BY

MWANA AUSTIN

TABLE OF CONTENTS

Introduction   4

Methodology   5

Failure Investigation   6

Material Description   6

Inspection and Tests   7-8

Operational Conditions of the heat exchanger   9

Effect of the Operational conditions   10

Effects of Corrosion   11-12

Detected Corrosion   13

Crevice Corrosion   13-14

Stress Cracking Corrosion   14

Pitting Corrosion   15-16

Conclusion   17

Recommendation   18

Appendix   19

References   20-21

INTRODUCTION

According to Aliya (2002), failure analysis is an investigative process carried out to determine the leading factors of an unwanted loss in functionality. The failure of structures, components or machines can be addressed, depending on the approach and base knowledge of materials engineering. Therefore, the application of failure investigation plays a key role in determining the root cause of failures. In carrying out failure investigation there are variables that guide in these assessment termed the 'HUFIEMOODS' this is used in evaluating the entire industry processes in order to reach the base cause of failure and design an effective solution.

In this failure investigation, we are going to be assessing the failure of a C22 tube bundle of a HP gas cooler heat exchanger. This heat exchanger is situated offshore and has undergone a series of repairs, servicing and materials upgrade within the last ten years, it currently has a tube leak after six months from its last repair. Our client Zonko Petroleum has supplied documents of defect assessments carried out by Eddy Current. We have also acquired further information from our client like; results of prior test analysis, thermal data sheet and the material analysis. This will serve as a guide in finding the root cause of failure and effective measures to rectify and prevent such failure.

METHODOLGY

To carry out a successful failure investigation, a sequence of measures must be covered.

- An analysis of all previous test is carried, to assess the result of these tests and efficiency of the provided solution.
- Assessment of the operation process and component of the heat exchanger. Note: sea water is used in the cooling effect of the heat exchanger, which deposits salt, Also high velocity and pressure flow around the walls of the heat exchanger.
- A material and design analysis of the heat exchanger to determine parts susceptible to different types of corrosion.
- A visual and internal inspection of the heat exchanger to detect other forms of corrosion and faults present in the heat exchanger.
- Information gathered from these process is then used to propose a better design or solution.

FAILURE INVESTIGATION

1.0   MATERIAL DESCRIPTION OF THE HEAT EXCHANGER

The tubing bundle was previously made up of C276 alloys but persistent corrosion lead to the use of C22 alloy. The C22 alloy is known for being resistant to oxidized and non – oxidized chemicals. It also protects materials from crevice attack, pitting and stress corrosion due to its well proven nickel – chromium – molybdenum components (Cieslak et al, 1986). The composition of C22 alloy is tabulated below based on Haynes online (2018).

Table 1. Material component

Components	C22 (weight %)
Nickel	57 balance
Cobalt	2.5 max
Chromium	20-22.5
Molybdenum	12.5-14.5
Iron	2.0-6.0
Tungsten	2.5-3.5
Manganese	0.5 max
Vanadium	0.35 max
Silicon	0.08 max
Carbon	0.01 max
Copper	-
Sulphur	0.02 max
phosphorous	0.02 max

Table 2. The mechanical properties at room temperature for the alloy sheets properties.

Property	C22
Tensile strength (ksi)	122
0.2 yield strength (ksi)	63
Elongation (%)	54
Hardness (HRb)	93

Assumption: It is assumed the alloy is a sheet.

2.0 INSPECTION AND TESTS

In proceeding with the failure investigation, tests and physical inspection of the heat exchanger is carried to determine the operating conditions of the system. **Visual inspection** – usually, the use of naked eye is first used to roughly evaluate the root causes and type of failure. From figure 1 and 2, there is a development of pitting corrosion in internal surface tubing and also from figure _, a development of crevice corrosion around the cladding region.

Figure 1: iron oxide deposit visible internal defect (case study, 2018)

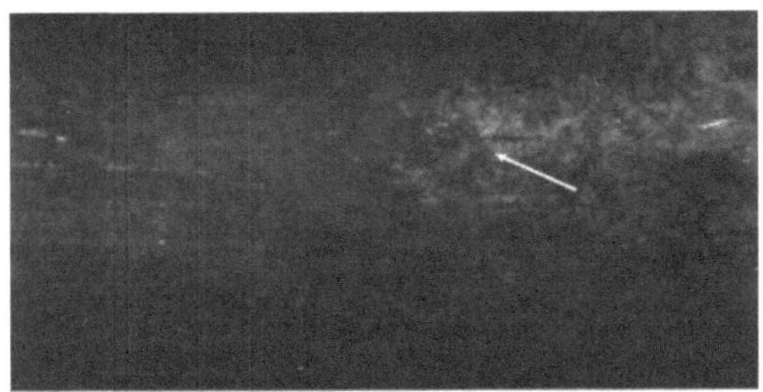

Figure 2: internal pitting underneath arrowed (case study, 2018)

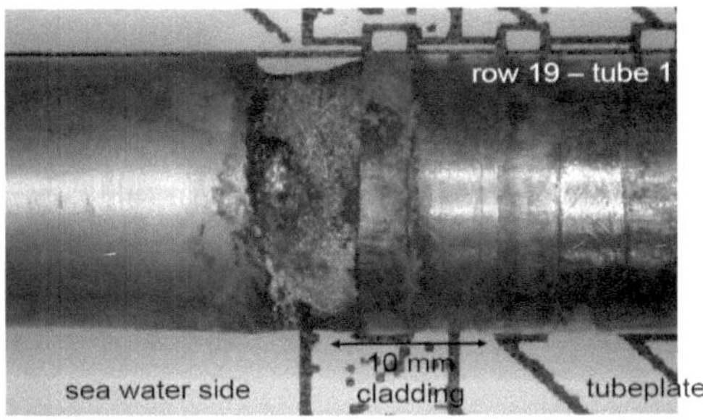

Figure 3: row 19 – tube r.h.s, typical example of a crevice corrosion

Cyclic passivation test – the test was carried out to study the behaviour of C22 alloys supplied by Haynes and Salem using the ASTM G5 under various conditions to measure the passivation behaviour. The result indicates that C22

supplied by Salem will remain passive even at a higher current and less surface area.

Crevice corrosion test - A variety of ASTM G48 classification C test with a critical pitting temperature up to 50⁰c for 72 hours was carried out to determine the better form of alloy material for the tubes to be used between C22 alloy and C276 alloy on the condition of it being used as built or over expanded. This qualifies the alloys possibility of crevice and pitting corrosion based on the nickel and chromium components at the specific temperature.

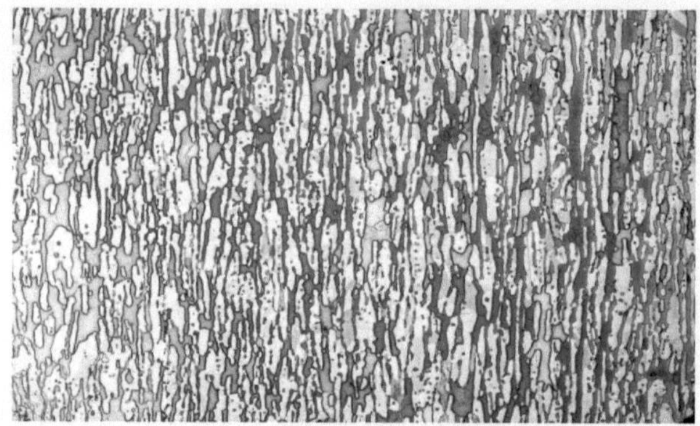

Figure 4: A correct etching of a stainless steel sample (G2MT, 2011)

From the results at appendix b, the change of the alloy will not aid in reducing the possibility of crevice corrosion occurrence because the as built condition of the C276 has the lowest maximum depth of crevice attack.

3.0 OPERATIONAL CONDITIONS OF THE HEAT EXCHANGER

Here are the corrosions resulting from the operation process: expansion and contraction from temperature variations, vibration from fluid flow, pressure on the tube walls from high velocity and pressure of the fluid.

From the schematic drawing of the C22 tube bundle part of the Heat exchanger, the designed and operating conditions in terms of temperature, at a rather high fluid velocity are tabulated below;

Table 3. Operational Conditions

Sea Water(Coolant)		
	Design	Operating temperature (^{0}C)
Cooling water in (^{0}C)	18	16.7
Cooling water out (^{0}C)	29.4	42.4
High Pressure Gas (HP Gas)		
	Design	Operating
High Pressure Gas in (^{0}C)	169	160
High Pressure Gas out (^{0}C)	49	40

3.1 EFFECT OF THE OPERATIONAL CONDITIONS ON THE HEAT EXCHANGER

3.1.1 Tubular Expansion

The removal of deteriorated tubes and reinstallation of newer ones in the shell has been carried out a number of times during the operational life of the heat exchanger in question, accumulatively reducing the integrity of tube to tube-joint connections. The Tube is held firmly in the shell baffle by the means of expansion of the tube by exertion of a force on the tubing material till its outer diameter is equal to the inner diameter of the shell baffles, and a little further more till the shell baffles area holding the tube reaches its yield point, repetitive implementation of this process will definitely lead to fatigue of the tube-plate joint hence creating a relatively larger clearance that can lead to escape or leakage of fluids. Like in Fig. 1 below.

The factors that determine choices of shell and tube types design pressure, temperature, differential movement between shell and tubes as well as fouling nature of the fluid. Technically speaking, all these factors affect the tubes affect they have been expanded, so all of this factors must be considered in the subsequent maintenance operation.

Fig 1. Leaks spotted on the tube bundle (C.W 2018)

This leak reduces the integrity and strength of the tube.

3.1.2 Effects of Corrosion

When a corrosion occurs, there are possible aftermaths which may not be seen alone by visual inspection. In this study; the C22 alloy being susceptible to corrosion although at a slower rate, the presence of untreated seawater which is

the North Sea has chlorine ion readily present and tensile stress in the form of residual stress due to heat treatment and plastic deformation occurring at the end of the heat exchanger (Evans et al, 2005). These are all recipe for a stress corrosion to occur. There is also a possibility of fouling due to accumulation of scales during shutdowns and lack of cleaning, this affects the heat transfer capacity and leads to variance in the heat exchanger design and operating temperatures.

3.1.3 DETECTED POINTS OF CORROSION

As part of the investigation, a visual inspection is carried on the material also corrosion tests are carried to determine isolated points of corrosion.

The spot where corrosion occurred can be spotted at the production gas inlet region which is not welded, exposed at high temperature and contain CO_2. Within this location there is an accumulation of scales due to lack of regular cleaning, these results to a poor heat transfer between the shell and tube contacts. Also because the tubes are not welded, a crevice was created between the over lapping tubing and shell metals which in an oxygen rich environment where the sea water acts as a bulk electrolyte, this will lead to crevice corrosion (Millero, 2002).

3.1.4 STRESS CORROSION AND CRACK (SCC)

The stress corrosion cracking as well as fatigue corrosion was due to repetitive cycles of thermal stresses resulting from the high operating temperature of the heat exchanger as well as the vibratory effect of the heat exchanger due to the flow velocity of the fluids in the exchanger. It is a rule of thumb that cracks are normally initiated at the stress concentration or localized corrosion points in the damaged are of the heat exchanger, in other words corrosion fatigue.

The expansion and contraction of the tube as a result of fluctuation of temperature reaching almost 170^0 can cause material fatigue and can be one of the causes of cracking, the cracking as a result can lead to leakage in the facility.

Fig 2. Cracking effects from stress and thermal stress corrosion

(C.W 2018)

4.0 DETECTED CORROSIONS

4.1   CREVICE CORROSION

This can take different forms depending on the materials alloy susceptibility and environmental conditions. Crevice corrosion commonly refers to deterioration of a metal happening in confined/small spaces. It is worthy to note that many heat exchanger failures are not from a single source but a combination of one condition leading to another. For example; crevice corrosion occurring as a result of pitting attack. This could be simultaneously at one location or at multiple locations.

CREVICE TEST

A variety of ASTM G48 classification C test with a critical pitting temperature up to 50^0c for 72 hours was carried out to determine the better form of alloy material for the tubes to be used between C22 alloy and C276 alloy on the condition of it being used as built or over expanded. This qualifies the alloys possibility of crevice and pitting corrosion based on the nickel and chromium components at the specific temperature.

CAUSES OF CREVICE CORROSION

    A. Presence of gaps/confined spaces gives rise to stagnation of fluids at the crevice and hence corrosion problem as a result of the following factors:

        i.   Improper sealants between the shell and the tubes

        ii.   Gasket failure

    B. Relative stagnant conditions of the fluids due to shut-in

    C. The presence of chloride concentrates in the fluids: ferrous ions form ferric chloride and attack on the crevices

    D. Expansion and compression of the tubes because of temperature effects

4.1.2 MECHANISM OF CREVICE CORROSION

This commonly occurs on metal to metal surface connection as shown in the figure 3.1a and b below. The propagation of crevice corrosion involves the depletion of oxygen that becomes localized on the outer surfaces close to the crevice, anode and cathode separation, dissolution of metal and metal ions hydrolysis *(Gordon, et al. 2005)*.

For effective crevice corrosion, there would be a crevice/gap wide enough to contain electrolyte and while providing stagnant conditions. The basic mechanism involving crevices corrosion in alloy is the exposure of chloride ions. This results to a gradual acidification of the solution inside the crevice leading to the destruction of the passivity and as such gives rise to an increase in the metal rate of corrosion.

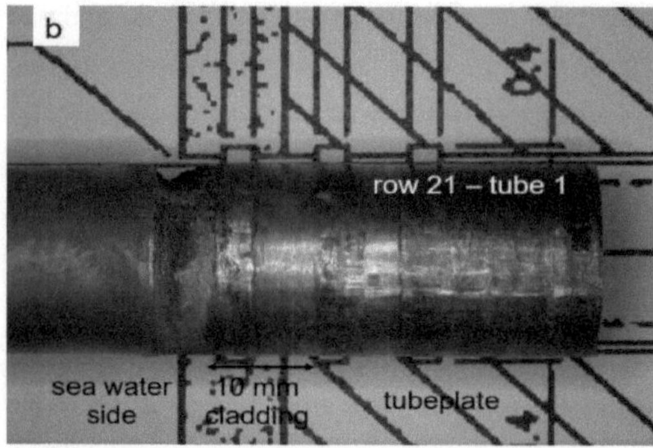

Fig. 3. Crevice corrosion

4.2   STRESS CORROSION CRACKING (SCC) PROPAGATION

The SCC and fatigue corrosion mechanisms were observed in the region of failure. It has always been thought that three conditions must be present simultaneously for SCC of the tubes to occur:

I.   A susceptible alloy

II.   Both static tensile loading and cyclic loading

III.   A critical corrosive environment *(Warke 2002)*

The destruction of the surface films layer was propagated by the action of fatigue loading and tensile stresses imposed on the tubes leading to Pitting or grain boundary corrosion. This spalling or localized corrosion give raises to SCC. The sequence of events involved in the SCC include three stages;

I.   The cracks initiation stage

II.   Steady state cracks propagation

III.   Cracks propagation/final failure

Fig. 3.2: Cracks on the tube. Row 19

4.2.1 CAUSES OF STRESS CRROSION CRACKING (SCC)

I.   Vibration: Cracks are initiated due to excessive vibration of the heat exchanger

II. Fluid velocity: in excess of 4ft/s can induce damaging vibration in the tubes causing a cutting action at support point with the baffles *(Schwartz 2010)*

III. Thermal fatigue: this normally happen at the U-bend due to accumulated stresses associated with repeated cycling

4.3   PITTING CORROSION

This localized form of corrosion occurred as a result of the deposition of water on the internal surface of the tubes. It is stimulated by the electrochemical reaction between the tubes internal surface and their environment.

4.3.1 CAUSES OF PITTING CORROSION

This is caused by the electro-potential difference between the tubing inside (anode) and the surrounding *(Jacobson 2010)*. This internal corrosion of the tubes occurred because of the direct interaction/reaction between the metal and the gas that flows through it depending on the following factors:

I.   High temperature: significant increase in temperature of the flowing fluid increases the rate of corrosion of the tube

II. Present of Impurities

III. Heat Exchanger shut-in period due to failure

4.3.2 MECHANISM OF PITTING CORROSION

The interaction between the anions and cations within the solution aid the formation of oxides film on the internal surface of the tubes as shown the 3.3a and b below. The iron oxidizes in the solution to form iron oxide and once this oxide is removed as a result of the factors listed above, thus, pitting corrosion emerge due to the exposure of the surface to the new gas flow *(Jonnes 1996).* Fig. 4.1 crevice corrosion (C.W 2018)

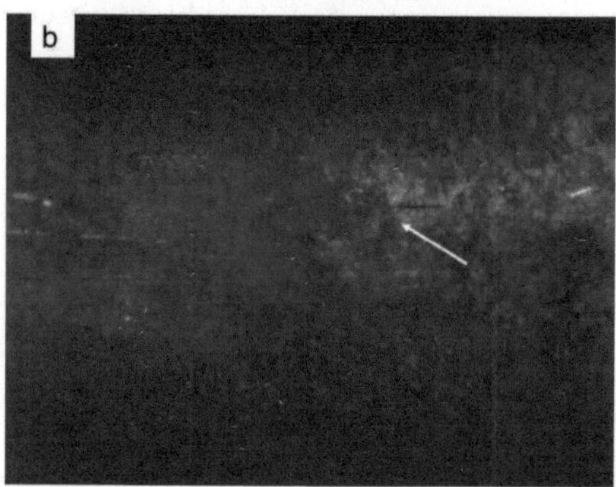

Fig. 4.2 Pitting Corrosion. (C.W 2018)

CONCLUSION

In conclusion, the failure investigation carried identifies corrosion as the main cause of failure. During the investigation different types of corrosion is seen to have occurred in main operating parts of the heat exchanger.

As we know, the operation of the heat exchanger depends on high temperature variations, resulting in the expansion and contraction of the inlet and outlet stream, while considering the thermal capacity of the material C22.

This expansion and contraction of the material led to the primary source of corrosion. The failure of the heat exchanger is not as a result of one type of corrosion but a simultaneous occurrence of different types of corrosion.

RECOMMENDATION

After an analysis of the causes of failure, it is our teams recommendation that the following measures be covered to prevent the occurrence of such failure:

MATERIAL SPECIFICATION AND DESIGN: Considering the operation of the heat exchanger, material connections should be considered to avoid the occurrence of corrosion. E.g. the use of API 660 in tube and plates assembly, this will reduce the chances of crevice corrosion and other corrosion process. Also a better welding or soldering option should be considered to reduce the chances of corrosion.

The heat exchanger operates in high velocity, high temperature, high pressure and heat transfer condition. These should be considered in the design of the baffle plate to prevent such failure from occurring again.

It is also recommended that the sea water used in the cooling effect should be treated to avoid the deposition of salt. Proper sealants should be used between the tube and the shell in order not to give way to crevice corrosion. There should also be routine inspection and replacement of faulty parts before the failure is magnified.

APPENDICE

Appendice A

East-Frigg Gas Pool Conditions	TEP/DP/LAB Composition Table 12-Report DV/cm/72		ADJUSTED COMPOSITION	
Element	% Weight	% Mol	% Weight	% Mol
CO_2	0,833	0,320	0,832	0,320
N_2	1,012	0,612	1,011	0,612
C_1	89,816	94,813+	89,724	94,812 (M
C_2	7,200	4,055	7,193	4,055
C_3	0,219	0,084	0,219	0,084
iC_4	0,036	0,010	0,036	0,010
nC_4	0,011	0,003	0,011	0,003
iC_5	0,015	0,004	0,015	0,004
nC_5	0,006	0,001	0,006	0,001
iC_6	0,027	0,005	0,027	0,005
nC_6	0,000	0,000	0,000	0,000
iC_7	0,017	0,003	0,017	0,003
nC_7	0,001	0,000	0,001	0,000
iC_8	0,075	0,011	0,075	0,011
nC_8	0,007	0,001	0,007	0,001
iC_9	0,025	0,003	0,025	0,003
nC_9	0,009	0,001	0,009	0,001
iC_{10}	0,056	0,007	0,056	0,007
nC_{10}	0,020	0,002	0,020	0,002
C_{11+}	0,615	0,063	0,716	0,066 (M)
TOTAL	100	99.998	100	100
MC_{11+}		165		

+ in fact the calculation gives :
CO_2 = 0,32053 and C_1 94,8137

(M) These values have been adjusted to obtain through calculation (SHG method) the working data.

Appendice B

In-situ ASTM G48-A (Modified) Crevice Corrosion Tests

Table 4.1 - 72Hrs @ 45degC
Result Summary

Test Block	Tube Ref	Distance of crevice attack*		Maximum Depth (mm)
		(mm)	(%)	
1 C22 as drawing	A1	0	0	0
	A2	0	0	0
2 C22 Over-expanded	A1	14.3	23.3	0.02
	A2	21.8	35.6	0.02
3 C276 As drawing	A1	0	0	0
	A2	0	0	0
4 C276 Over-expanded	A1	57.1	93.2	0.035
	A2	48.1	78.5	0.02

* Distance around circumference of tube (OD = 19.5mm)

Table 4.2 - 72hrs @ 50degC
Result Summary

Test Block	Tube Ref	Distance of crevice attack*		Maximum Depth (mm)
		(mm)	(%)	
1 C22 as drawing	B1	61.3	100	0.22
	B2	8.2	13.4	0.025
	B3	0	0	0
2 C22 Over-expanded	B1	54.2	88.5	0.015
	B2	26.0	42.4	0.02
	B3	49.0	80.0	0.012
3 C276 As drawing	B1	0.5	0.8	0.02
	B2	0	0	0
	B3	0	0	0
4 C276 Over-expanded	B1	58	94.7	0.025
	B2	61.3	100	0.075
	B3	61.3	100	0.04

* Distance around circumference of tube (OD = 19.5mm

REFERENCES

ALIYA, D., 2002. The failure analysis process: an overview. *Materials Park, OH: ASM International, 2002.*, pp.315-323.

CIESLAK, M.J., HEADLEY, T.J. AND ROMIG, A.D., 1986. The welding metallurgy of HASTELLOY alloys C-4, C-22, and C-276. Metallurgical Transactions A, 17(11), pp.2035-2047.

EVANS, K.J., YILMAZ, A., DAY, S.D., WONG, L.L., ESTILL, J.C. AND REBAK, R.B., 2005. Using electrochemical methods to determine alloy 22's crevice corrosion repassivation potential. *Jom*, *57*(1), pp.56-61.

G2MT Labs., 2011, "G2MT Labs-Cost of Corrosion Study," June 1, 2011, accessed Apr. 25, 2018, from http://www.g2mtlabs.com/2011/06/nace-costof-corrosion-study-update.

Haynes Online Literature: Hastelloy C-22HS Alloy at a Glance, accessed April 26, 2018. filename: H-2120.pdf. <www.haynesintl.com>.

MILLERO, F.J., 2002. Sea water as an electrolyte. In *Chemistry of Marine Water and Sediments* (pp. 3-34). Springer, Berlin, Heidelberg.

OWEN, J., 2018. Case study ENM 233(Group Coursework). Material and corrosion. The Robert Gordon University Aberdeen, Department of engineering, Sir Ian Wood building, delivered 13 April 2018.

Total Oil Marine Limited., 1977. Frigg field project. Design manual – volume 1.

BETTS, A. J. AND BOLTON, I.H., 2013. *Crevice Corrosion: Review of Mechanism, Modelling, And Mitigation.*

GORDON, M. G. AND MON G. K., ET AL., 2005. *Stifling of Crevice Corrosion. In Alloy 22.* In: T. R. Allen, P. J. King and L. Nelson Eds. The Minerals, Metals and

Material Society. [Online]. Available At:
Http://iweb.tms.org/NM/environdegxii/1431.pdf. Accessed On 25 April 2018.

JONES, R. E., 1992. *Mechanism of stress corrosion cracking.*

JONNES, DENNY A., 1996. *Principle and Prevention of Corrosion.* 2nd ed. NJ:
Prentice-Hall, Inc.
MARVIN, SCHWARTZ J., 2010. *Four types of heat exchanger failures.*

WARKE, R. W., 2002. *Stress Corrosion Cracking.* In: W. T. Baker and R. J.
Shipley Eds. ASM Handbook. Vol 11. pp 850 [Online]. Available At:
www.knovel.com.

YOUR KNOWLEDGE HAS VALUE

- We will publish your bachelor's and
  master's thesis, essays and papers

- Your own eBook and book -
  sold worldwide in all relevant shops

- Earn money with each sale

Upload your text at www.GRIN.com
and publish for free